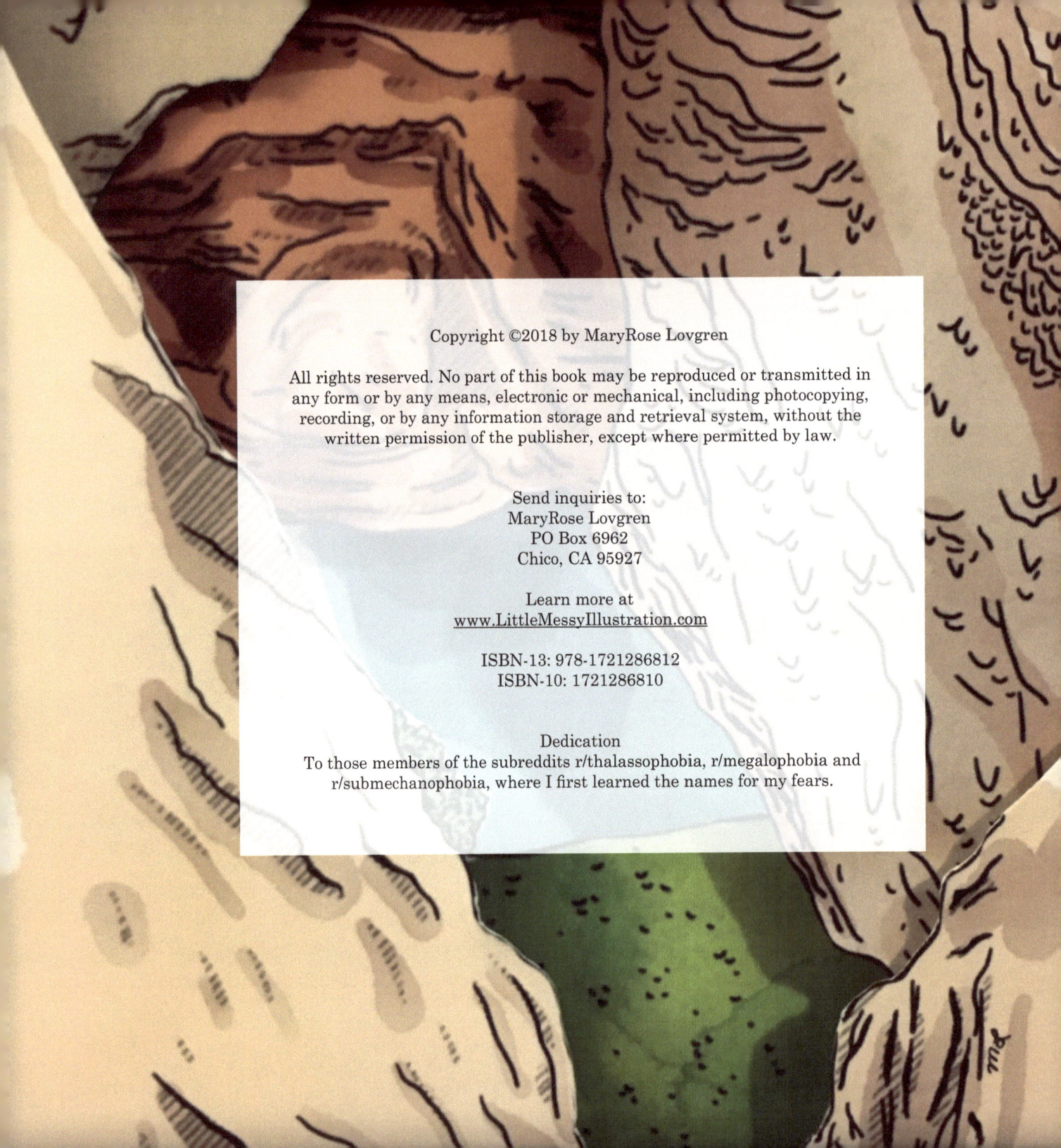

Copyright ©2018 by MaryRose Lovgren

All rights reserved. No part of this book may be reproduced or transmitted in any form or by any means, electronic or mechanical, including photocopying, recording, or by any information storage and retrieval system, without the written permission of the publisher, except where permitted by law.

Send inquiries to:
MaryRose Lovgren
PO Box 6962
Chico, CA 95927

Learn more at
www.LittleMessyIllustration.com

ISBN-13: 978-1721286812
ISBN-10: 1721286810

Dedication
To those members of the subreddits r/thalassophobia, r/megalophobia and r/submechanophobia, where I first learned the names for my fears.

SCARY PLACES,
Illustrated

Paper dioramas of bottomless pits, oversized structures
and all manner of natural and unnatural scary places.

by M.R. Lovgren

Introduction

Do you find deep-water caves, drains in lakes and ridiculously deep swimming pools a little...scary?

I know I do. But I'm also strangely compelled to confront these fears by lurking in online groups devoted to phobias of such things. Think submerged objects, oversized structures and, of course, the ocean.

The photos shared in these groups enthralled and frightened me to the point that I felt compelled to draw them. To fully immerse myself in my fears, I decided to make them 3D. I printed, cut and pasted each drawn layer onto foam core and placed the resulting diorama into a cardboard frame.

Photographs of these dioramas span the pages of this book. To complete the experience, I paired each photo with a detailed and well-researched description of each place. Some are of man-made structures, while others occur naturally through quirks of geology. But all of them scare me on a deeply primal level.

I hope you "enjoy" viewing these pieces as much as I did in creating them.

-M.R. Lovgren

Opposite page: A collection of the paper dioramas, with some "cozy places" thrown in.

  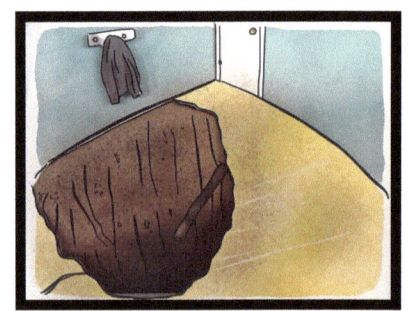

The Scary Places

DEVIL'S HOLE FRESHWATER CAVE   8-9
Y-40 DEEP JOY   10-11
BEDROOM SINKHOLE   12-13

TOKYO FLOOD CONTROL SURGE TANK   14-15
MIR DIAMOND MINE   16-17
THE STRID   18-19

BELL-MOUTH SPILLWAY   20-21
ELEPHANT'S FOOT, CHERNOBYL   22-23
KOLA SUPERDEEP BOREHOLE   24-25

DEVIL'S HOLE
FRESHWATER CAVE

Imagine taking a walk out in the hot desert of Nevada. The rocky dirt trail crunches underneath you. The sun is bright and hot and reflects off clumps of hard rock so that you have to shield your eyes with a flattened hand. All around you everything is beige, the color interrupted only by sage green bushes scattered across the valley floor and the grey-blue mountains rising up beyond.

Scrambling down some boulders in a narrow gully you come across a hole in the ground, a small rectangular pool of clear, deep blue water. There is a short algae-covered shelf of rock jutting out across a third of the pool, so you roll up your jeans and step in. The water is warm and welcoming, and as you slosh forward, you get to the shelf's end. Peering over the side you see shimmering shelves of blue and white and green. You decide to take a dip. As you ease yourself into the water, you find a need to tread water, as the sides of the pool are vertical rock, and your feet don't touch the bottom of this narrow oasis.

Fact is, your feet won't touch for at least another 400 feet down. You are now dangling on the surface of a column of dark water over 40 stories high.

Devil's Hole is the entrance to a vast limestone chasm whose depth is still not known. United States Geological Survey cave divers sank through the inky blackness to the 436 feet mark in 1991, and with their powerful flashlights could still see for another 150 feet below before it curved out of sight. According to professional diver Jim Houtz, "at the end of the tube it opens again into something else. We don't know what the next room is, or if it's a room at all. It's like infinity."

The unique nature of Devil's Hole allows it to perform another cool trick: It serves as an indicator of distant seismic events. An earthquake occurring thousands of miles away will shake the vast reservoir of water deep within Devil's Hole, and having nowhere to go but up, the pool will violently climb the rocky sides of the cavern only to be sucked down again. This occurs over and over again, appearing very much like a toilet being flushed.

Unfortunately for those enticed by bottomless pits and the claustrophobic abyss, Devil's Hole is off limits to the public, as it houses one more secret: It is home to the tiny, critically endangered Devil's Hole Pupfish. In fact, it is the only home of these little bluegrey (male) and brown (female) fish, whose total population hovers around one hundred. These pupfish, or C. diabolis, are extremely endangered, and so their sensitive habitat is surrounded by a tall prison-like barbed wire fence. You can walk out along an enclosed walkway to view the hole from above, but that is as close to the bottomless void as you will be able to get.

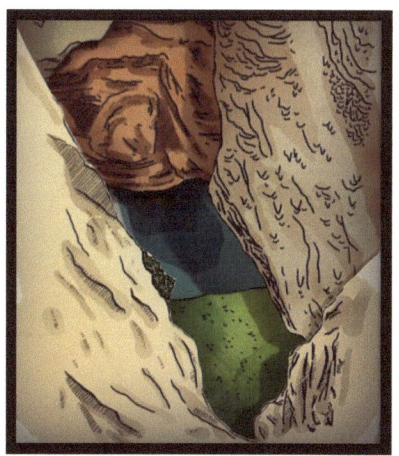

Opposite page: Devil's Hole from below.
This page: Devil's Hole from the surface.

Y-40
DEEP JOY

My hometown high school had a large swimming pool that was open to the public each summer. It was shaped like an L and had three landings at different depths: It started off at four feet, gently sloped down to six feet, and finally, below a high diving board, sank to a final depth of thirteen feet.

As a kid, I would hold my breath and dive down to the bottom of the six foot section and let my eyes wander slowly over to the darkness of the deep end. Though weightless, catching a glimpse of that abyss made my stomach drop. You couldn't see the very bottom until you had crept along to the very edge of the six-foot drop off, which of course I would do repeatedly. It was only then that you saw an added attraction: the presence of a very large drain at the bottom with a grate covering it.

When I was a little older, I would hold my breath, and, if I swam hard enough, dive to the very bottom. On days I was feeling particularly brave (or foolhardy), I would briefly touch the grate and then race for dear life back up to the surface.

I recently saw a video of a swimming pool that is quite a bit deeper. Called the Y-40 Deep Joy, it has a final resting place of 40 meters (130 feet). That's ten times (10! times!) deeper than the pool of my childhood. In fact, the Deep Joy is currently the deepest indoor swimming pool in the world (with a Guinness World Record to prove it). Located in a hotel in Padua, Italy, it features underwater caves for divers to explore and, like my school's pool, platforms at different levels, including one at 4.3 feet and another at (gulp) 39 feet. There is a clear tunnel placed just under the water's surface for visitors to watch scuba and free-divers dance in the deep. And of course, there is the deep end, which take the form of a narrow cylindrical well that plunges in semi-darkness to the very bottom.

Search up this pool on the Internet and you will find a video of French free diver Guillaume Nery smoothly swimming through the depths of the pool. This video teases the presence of the bottom in brief snippets as Nery swims from platform to platform. It is only when he reaches the edge of the well that the camera finally faces it head on, allowing the abyss to fill the screen. Nery drops into this darkness like a ballet dancer, arms crossed over his chest, feet together pointing down. His eyes are closed and he looks strangely peaceful as he sinks to the bottom.

You may find yourself holding your breath with him. You might also find yourself watching the video over and over, scaring yourself anew each time. Until, of course, they build a deeper pool.

BEDROOM SINKHOLE

One night in February, 2013 after Jeff Bush had gone to sleep, a sinkhole opened up under his bed. The rest of his family were still awake when they heard the collapse, comparing the sound later to a car crashing into their house.

Jeff's brother, Jeremy, raced to Jeff's room after hearing him call out, but all he found was a hole that had swallowed Jeff's bed, his dresser and his TV. Only the cable wire dangled from the wall into the darkness.

Sinkholes are nature's dirtiest trick. It takes the ground beneath our feet and pulls it out from under us so that we can fully inspect the horrifying void while simultaneously falling into it. Sinkholes are also sneaky. They usually develop under the earth silently for years and years until the ground above can no longer support itself, and the unlucky thing sitting on top is rudely dropped into an abyss.

Sinkholes often form where rock such as limestone is in abundance, as this rock dissolves in water over time. In states like Florida (which is built on limestone with a veneer of clay on top), sinkholes are so ubiquitous that insurance companies are required to provide homeowners insurance coverage that includes damage from "catastrophic ground cover collapse."

Catastrophic ground cover collapse is one of the things I like to think about when I go to sleep at night.

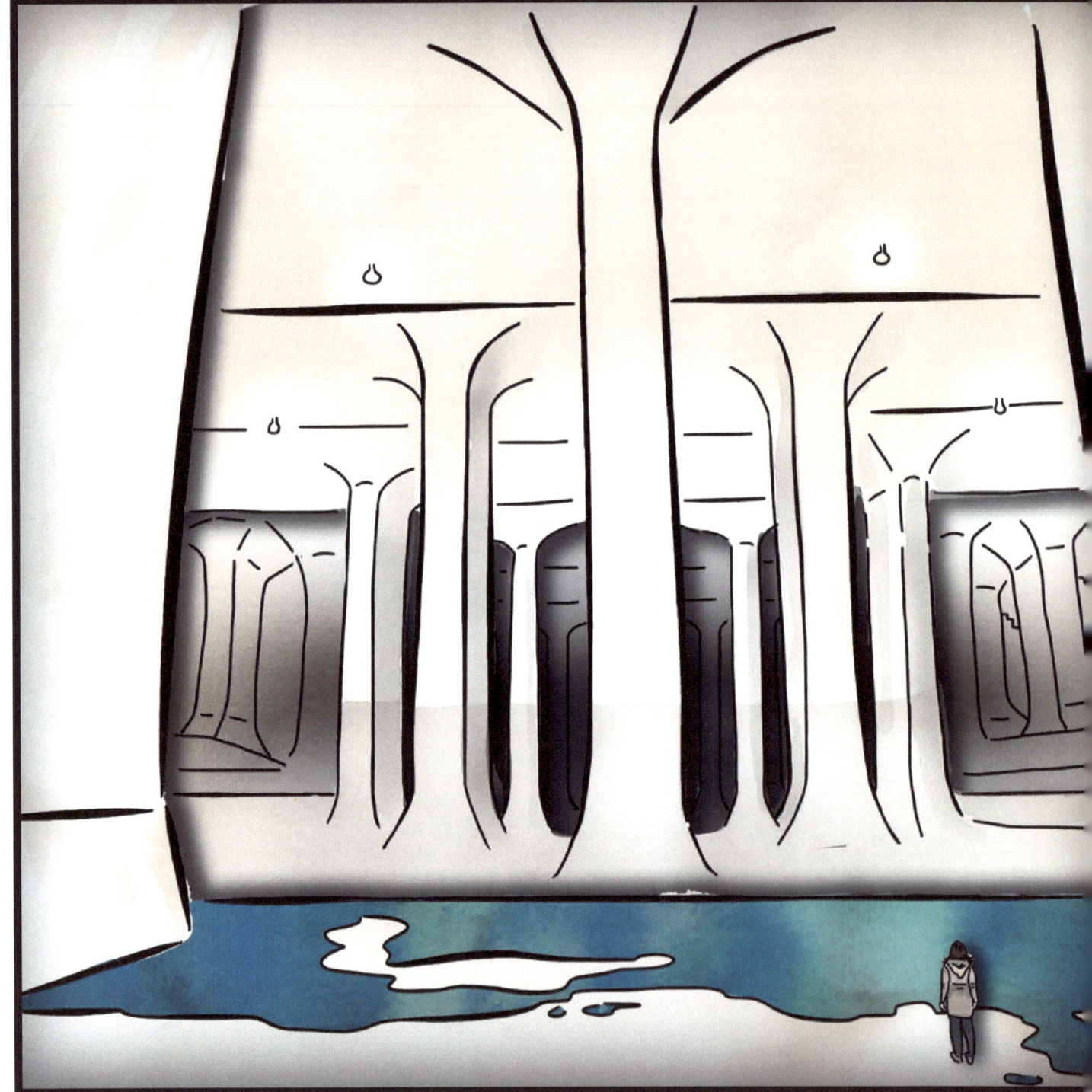

TOKYO FLOOD CONTROL SURGE TANK

Huge columns rise into the darkness, resembling smooth stalagmites in an enormous subterranean cavern. The floors are smooth as well and resemble marble from a distance. Sections of the pillars are highlighted in the darkness by lights set up at intervals. The combination of the columns, the lighting and the vastness of the space inside lends it an ethereal quality, like that of a temple buried deep underground.

But the pools of water on the floor give a subtle indication of the true purpose of the sacred space—to serve as a massive flood control channel in an area prone to flooding just outside of Tokyo, Japan.

That's right, this place fills with water. A lot of water.

A number of rivers criss-cross the city, and when those rivers rise during monsoon season (or even just during heavy rains), the streets of Tokyo begin to flood. With the creation of the Metropolitan Area Outer Underground Discharge Channel, that water is instead diverted underground into the nearest of five vertical silos throughout the city, which then connect to a series of underground tunnels almost 4 miles long.

If the water in the tunnels and the silos rise too high, it flows into this place, the largest surge tank in the world.

And large it is. Each pillar holding up the ceiling of this massive tank is almost 60 feet high. The room itself is 580 feet long, or the length of two football fields. Overflowing flood water fills this space anywhere from 7 to 12 times a year, and then is pumped out into the Edo River.

When it isn't full of water, it is a very popular tourist attraction. You read that correctly—you can climb down into this surge tank and imagine what it's like when it's full. Or, stick around and see for yourself.

MIR DIAMOND MINE

The Mir Diamond Mine is the largest open-pit diamond mine in the world. At 1,722 feet deep and 3,900 feet wide, it is also one of the largest excavated holes in the world. I stare at pictures of this mine often because the sheer size of it boggles my mind. The little tiny boxes scattered around the surrounding land are actually buildings several stories high. You can barely make out the cars, let alone any people.

This mine is found in the Siberian region of Russia in the town of Mirny. Development started in 1957, and if the word "Siberia" means anything to you, you can imagine the challenges that were ahead. Winter there lasts seven months of the year during which the ground is frozen. And when I say frozen, I mean car tires would freeze and steel would shatter. Jet engines had to be brought in to thaw the ground to dig. In the short summer months, however, the ground softened and buildings would sink into the permafrost, necessitating that they be built up on piles.

You might ask yourself why anyone would spend so much effort digging such a big hole in Siberia. Did I mention that this is a diamond mine? At its peak, 10 million carats of diamonds were produced each year. This was enough to worry another company who dug for diamonds, De Beers. In fact, De Beers had to buy diamonds from the Mir Mine to control the market price of the gem.

But this wasn't enough— they wanted to know more about Russian mining developments. In 1976 Russia acquiesced to their plea to visit the mine, and an executive and the chief geologist of De Beers were granted visas to check it out for themselves.

Once the pair landed in Russia, however, they found themselves forced to attend party after party until they were seriously worried that they would never get to see the mine. They were finally granted a short visit just before their visas expired. When I say a short visit, I mean they had literally 20 minutes. Well played, Russia.

The big hole is no longer mined for diamonds. But get this: Now they mine under the hole. Up to 1 km beneath it. That may actually be worse.

THE STRID

There is a waterway in Yorkshire, England that appears very much like a pleasant mountain stream. Rocks made soft with green moss crowd the edge and white water swirls and drops from the many small waterfalls along its length. Only six feet across is most places, you might be tempted to try to playfully jump across while picnicking with friends.

But this would be a terrible mistake, because it isn't a stream at all.

The Strid, as it is called, is actually a river. A river turned on its side.

Upstream of the Strid you will start to understand. The water coursing its way through this false brook is actually the River Wharfe. Here it acts more like rivers you have known, expanding to a luxurious 30 -40 feet across. But at the Strid it is suddenly constricted by geology, and instead of spreading out, it has had to spread down. The "brook" at the top hides an immense and deep canyon filled with water. Over time the river has carved out channels and chasms under the rocks above. Any living thing that falls in is immediately dragged down in a vortex of currents and then thrashed and pummeled and ultimately trapped within its undercut banks.

There is a claim that no one has ever survived a fall into the Strid, and while that is difficult to substantiate, there certainly have been many victims, one of them a possible future King of Scotland.

In 1154 a young boy named William de Romilly decided to make the leap across only to fall in and disappear. His mother, Lady Alice de Romilly, was so beset with grief that she donated the land around it to the Bolton Priory monastery. William Wordsworth later immortalized the tale in his poem, "The Force of Prayer."

The Boy is in the arms of the Wharfe,
And strangled by a merciless force;
For never more was young Romilly seen
Till he rose a lifeless corpse.

BELL-MOUTH SPILLWAY

The idea behind a spillway is simple: Prevent water from rising too high in a reservoir. Much like the overflow tube in your sink or bathtub, the spillway of a dam allows water that has reached a critical height to be diverted into an enormous drain to be released into the river below. A Bell-Mouth Spillway, so named due to its shape, is unique in that it is placed within the reservoir itself. In dry periods, it merely looks like a ring-shaped column of cement protruding out of the water. But all of this changes when the reservoir starts to fill.

Once water levels rise high enough to reach its top, this type of spillway actually disappears. The water flowing over the edge is the only indication of its presence. It becomes a black hole in the midst of tranquil waters. If this occurs near a roadway, you will often see onlookers crowding near to see it in action. But the inner workings of this type of spillway are hidden deep within the black depths of the dam. You can't see where it goes, but you know it's deep.

Another part of the horror of this type of spillway comes from its uncontrolled operation. There is no mechanism to prevent it from starting. Once water levels reach its top, the water goes in. No technician turns it on, no gate prevents objects from entering. There is no intelligence at work here; it is simply a giant drain in a lake.

At least one person has been pulled into a Bell-Mouth Spillway and drowned. In 1997, a woman succumbed to the so-called "Glory Hole" in Lake Berryessa, in Napa, California. One account says she was fully clothed and distraught when she entered the lake. She then either swam or was pulled toward the opening of the spillway. As she was swept up and over, she grabbed onto the concrete ledge and clung there for twenty minutes, slowly losing her strength as the power of the water flowed around her. Onlookers could only watch until, finally, she disappeared inside. Her body was found in the river below, having made the journey completely under the dam and out the spillway's exit.

Opposite page: The opening of a Bell-Mouth Spillway. This page: View of opening and exit of spillway in Lake Berryessa, California reservoir.

ELEPHANT'S FOOT, CHERNOBYL

If anything can be accurately described as a "Hot Mess," it's the Elephant's Foot, an oozing pile of radioactive waste deep in the basement of failed Reactor Number 4 at the Chernobyl Nuclear Power Plant. Often referred to as the deadliest object in the world, merely standing in the same room with the Elephant's Foot for five minutes would kill you in two days.

The one way I don't want to go is from radiation poisoning. Intense doses of these invisible rays make your cells fall apart. That's not good. You basically slough apart from the inside out. And it hurts. Cue lots of vomiting and skin melting. The word "hemorrhage" gets thrown around a lot.

But back to the Elephant's Foot.

When the Chernobyl plant blew its top in 1986, the nuclear reactor core got so hot that it literally melted down. It melted the steel and concrete containing it and flowed down and out of the bottom of the reactor vessel like the worst kind of lava. It pooled in the basement below as a new substance nicknamed "corium."

When it was discovered months after the disaster, its blobby shape resembled the foot of an elephant and this unique sample of corium got its name. While no longer as hot as it was over thirty years ago, you still don't want to visit it. Down from its peak radiation of 10,000 roentgens per hour (or how much radiation you would get from 4.5 million chest x-rays), it will still kill you if you spend over an hour with it. And it's still melting.

Of the 600,000 "liquidators" who worked at the Chernobyl site after the disaster to contain the mess, more than 30 died within a few months of radiation poisoning. We owe a lot to all of these people for helping to contain one of the worst human-caused disasters of all time.

KOLA SUPERDEEP BOREHOLE

The deepest hole in the world is under this welded-shut cap in an abandoned structure in Russia. Called the Kola Superdeep Borehole, this 9-inch diameter hole goes deeper than the bottom of the sea.

Read that again: This hole goes deeper than then lowest known point in the ocean. Challenger Deep, at the bottom of the Mariana Trench, is 36,201 feet below sea level. The Kola Borehole goes over 4,000 feet deeper.

All of this of course begs the question: Why dig such a deep hole? Russian scientists hoped to learn more about the continental crust, which is the layer of rock that forms the continents and their continental shelves. They made it 7.6 miles, or about a third of the way through the Baltic Shield crust. (Keep in mind that the distance to the center of the Earth is 3,958 miles, so in comparison, it barely pricked the surface.)

Drilling began in 1970. A 200-foot tall superstructure was built to house the enormous pipes that made up the "drill string" which connected the drill to the surface. As the drill reached lower and lower, sections of pipe were strung together like the world's longest necklace and then lowered into the hole. In 1984, a 3-mile section of this drill string twisted off and ended up blocking the hole. They had to start again with a new hole four miles down.

Some of the cool things that they discovered included water below an impermeable layer of rock (formed from oxygen and hydrogen molecules being literally squeezed out of minerals) and a high quantity of hydrogen gas. They also discovered fully intact fossils of microscopic single-cell plankton, over 2 billion years old!

In 1992 drilling finally stopped, as they encountered higher-than-expected temperatures at this depth (up to 356 degrees Fahrenheit).

With the end of the Soviet Union in 1995 came the abandonment of the site. Garbage is now strewn across one of the greatest human achievements of the 20th Century.

If you do ever visit the Kola Borehole, please don't open it up. You wouldn't want to drop anything important down there.

About

MaryRose Lovgren has degrees in Zoology and English Literature from the University of California at Davis and uses her background in science and storytelling in her art. She has a fondness for pen-and-ink drawing and the illustrations in children's storybooks. Her art and writing have been published in online journals and magazines.

You can view more of her work at
www.LittleMessyIllustration.com

www.ingramcontent.com/pod-product-compliance
Lightning Source LLC
Chambersburg PA
CBHW042322250526

R18347300001B/R183473PG45473CBX00014B/3